I0491016

About the Author

Sam Puleo is the owner of Shwego Media. His company has helped many plumbing businesses increase their revenues and profits through the implementation of internet market strategies. In this book, Sam details the specific strategies that have worked for him and his plumbing clients, so you can implement them in your business.

This book is dedicated to my loving wife, Anita, and our daughter who is on the way – Anna Nora. I love you.

"We had some horrible experiences with some of the bigger "brand name" SEO and Marketing Companies so we were very distrusting. Sam and the team at Shwego changed my mind completely. They have taken our business from the basement of the Google map listing to competing with the large companies in our area in less than 6 months"

-Pro N Stall

"We partnered up with Shwego Media for them to handle our marketing on Google and social media. Sam is awesome to work with, always responds to you stays on top of everything. Keeps you posted on how campaigns are performing. I highly recommend these guys, they definitely know their stuff."

-ASAP Plumbing Services

"Shwego has been excellent since day one! They are a very valuable addition to our business and we would highly recommend them to anyone looking for marketing. Sam and Rich have been super helpful every step of the way and are able to thoroughly explain everything to us. Again, we highly recommend Shwego for your marketing needs!"

-bluefrog Plumbing, Connecticut

"I knew we made the right decision when I spoke with a client on the phone who turned out to be the marketing manager for a very large home building company in our area. She said, 'You guys have an awesome web presence! I'm very impressed with what I saw when I looked up your company!' We have gotten several other compliments from competitors in our area. Thank you Shwego for taking our company's success personal!"

-bluefrog Plumbing, Houston

"Shwego is the best! Great service at a great price. We've never been happier with our digital marketing partnership. We highly recommended Shwego!"

-bluefrog Plumbing, California

Introduction

If you're reading this book, I will assume that you own a plumbing business and are interested in growing your business using modern technology and online marketing strategies.

My name is Sam Puleo and I am the owner of Shwego Media. We specialize in helping plumbers acquire new customers online, and this book will teach you all of our methods so you can do them yourself.

This book is written to be used as a guide, so feel free to jump to the sections that you think apply most to your business. All sections are outlined on the next page.

If you have ANY questions about the topics in this book, I would encourage you to contact me directly. I am happy to answer your questions and further explain these concepts to you in detail.

My email address is sam@shwego.com

Thanks so much for your interest, and I hope this book is a big help to you as you grow your plumbing business!

Contents

- **Overview of Online Customer Acquisition**
  - Understanding the internet landscape
  - Setting goals and developing a plan
  - Digital customer acquisition costs

- **Local Google Maps Rankings**
  - How Google ranks local plumbers
  - Ways to increase your ranking
  - Ongoing tasks to continually improve

- **Google Ads**
  - Google Local Services Ads (Google Guaranteed)
  - Google Pay-Per-Click Ads (PPC)
  - Google Local Campaigns

- **Your Website**
  - How to develop a quality website
  - Why your website content matters
  - How to get your website ranking higher

- **Ongoing Marketing**
  - The importance of retention and referrals
  - Ways to build a customer following
  - How to set up ongoing marketing campaigns

- **Technology**
  - Implementing a CRM
  - Integrating marketing into your CRM
  - Using technology to scale your plumbing business

Customer Acquisition

Customer acquisition, or simply put getting new customers, is critical to the success of any business. In plumbing, you must have a steady stream of new customers to pay your bills and earn a healthy profit. Repeat customers and referrals are always nice, and we will teach you how to create more of them later in this book; however, a plumbing business that isn't constantly getting new customers will fail quickly.

There are many ways to acquire new plumbing customers online. Let's discuss them in detail below...

Lead Generation Platforms

Websites like Yelp, Angi, and HomeAdvisor offer customer leads to plumbers through their sites. They spend millions of dollars on marketing to get their websites in front of customers online and you can buy customer leads from these companies.

- Positives: you can get the phone ringing right away.
- Negatives: the leads are costly and customers are being sent to you and your competitors.

Google

In my opinion, Google is still the mothership where most people go when they have a plumbing issue and need to find a reputable plumber. These people have true buying intent and if you can find ways to get in front of them, your phone will ring. There are a few ways to get in front of customers on Google:

- Google Local Services Ads
- Google Pay-Per-Click Ads
- Google Maps
- Google Web Search Results

We will discuss each of these in detail in the chapters to come and show you how you can effectively get in front of new customers on Google.

Making a Plan

The first thing you must do before you venture into any online marketing or advertising is to make a plan. You need to know the types of customers you would like to acquire, the services you want to market most, your average profit margins on those services, and your total budget. On the next page, there is a worksheet for developing a profitable online marketing plan.

Planning Worksheet

Where Are Your Customers Located?

_____

What Services Do You Want To Market Most?

_____

What Is Your Average Profit Margin Per Job?

_____

How Much Money Do You Have To Invest?

_____

Although these questions won't give you all the answers, they are a great place to start. If you're planning on marketing drain cleaning services, and the average drain cleaning job makes you $80 in profit after materials, labor, and overhead, you cannot afford to purchase leads online for $100 or you will lose money... even if the phone is ringing more!

The biggest mistake I see plumbers make is not knowing their true numbers and pouring money into lead generation. This is the fastest way to go out of business.

When you know your true profit margins on each service performed, you can develop a better marketing plan.

In the example on the previous page, drain cleaning might not be a great service to focus on when advertising due to the slim margins. Tankless water heater installations, however, might produce larger margins where you can afford to buy leads or spend money on Google Ads.

The first thing I do with all of my plumbing clients is to make a financial plan, so we understand where our money is going and our true return on investment.

Customer Acquisition Cost

Your next job after understanding your true profit margins on each service is to calculate customer acquisition costs. Depending on the platform you are using, these costs can vary widely. Your costs on things such as the Google Map and Google Search results can simply be the time you invest or pay someone to complete the tasks necessary to get your business ranking.

Other platforms, such as Google PPC Ads, Yelp, or HomeAdvisor, will have a true cost to acquire a customer.

Customer acquisition cost is fluid and should be calculated regularly. I encourage my clients to calculate customer acquisition costs by online channel each month. Here is an example below:

- You spent $2,000 on Google PPC Ads
- You received 100 phone calls
- Those calls resulted in 20 paying jobs

Customer Acquisition Cost = $100 (you got 20 new customers for a total cost of $2,000)

Another metric that is important to track is your Cost Per Lead or Cost Per Phone Call. In the example above, your total cost per lead is $20 (you received 100 calls for $2,000).

These metrics are important to track and monitor. The ultimate goal is to continually decrease customer acquisition costs across all online channels.

Some of our clients' CRM systems allow us to integrate marketing campaigns. This allows our clients to assign revenue based on the channel from where the customers came. In the example above, you would be able to determine how much revenue your $2,000 investment produced. This takes some technical skill and it is best to hire a company to do this for you.

Google Maps

The Google Map is currently the number one way to acquire customers online for plumbers. The map was designed by Google to allow searchers an easy way to find a local business, and many phone calls can be generated from a high map ranking. In addition, ranking in the map has nothing to do with advertising, so if you can secure a high position by doing the right things, your customer acquisition costs are much lower.

How Google Ranks The Map

To truly understand the impact on your local map rankings, you first need to understand what Google is trying to accomplish. Google's customer is the searcher, and Google is trying to give their customers the best experience possible. If someone is looking for a plumber and the top plumber on the map is not reputable, that person might hire that plumber and have a bad experience. Google is concerned that the customer will blame their experience on Google and might not come back. With that said, Google is trying to rank the most reputable companies at the top of the map.

Since Google is technologically-driven and they do not have the manpower to individually evaluate every single business in the country, they rely on their computer algorithms to determine who the most reputable business are that deserve a higher map ranking. Here are the elements that determine how you rank on the map and why...

Customer Reviews

Currently, customer reviews carry the most weight when it comes to ranking your plumbing business higher on the Google map. Since Google is trying to display the most reputable plumbers to the searcher, one way of determining your credibility is through your customer reviews. Here are the aspects of customer reviews that matter most to your ranking:

- Review Velocity: how often you get a new customer review and how quickly the number of reviews you get are increasing - this shows Google that you are an active, growing business. If you are growing, you must be doing something right.

- Total Review Count: how many reviews you have on your profile gives Google an indication of how many jobs you have completed. The more reviews you have, the higher you rank.

- Review Content: what people are saying in your reviews is also very important when ranking for specific searches. When customers mention how well you installed a water heater, this helps you rank higher for water heater searches.

- Review Rating: your total review score on a scale of 1-5 shows Google how happy customers are with your service. The higher the rating, the higher the ranking.

- Review Responses: responding to the customers who leave you reviews is extremely important. Not only is it good practice to thank your customers, but it also shows Google that you care about customer service. In addition, when people are looking at your business on Google and they see that you are thanking every customer, they might be more apt to call you for service.

Business Profile

The next important aspect of getting your business to rank higher on the Google map is to complete your Google My Business profile according to Google's latest standards. There are many aspects of the GMB profile that you need to address.

Here are the most important things to complete on your GMB profile:

- Name, Address, Phone Number (NAP): make sure your business name address and phone number is listed exactly how it appears on other business listings such as Yelp. When Google is able to identify your business on other websites through an identical match of your NAP it helps increase your credibility and ranking on the map.

- Service Options: select the service options that are relevant to your business.

- Hours of Operation: be sure to list your business hours properly and update your Holiday hours as the Holidays approach.

- Appointment Link: be sure to include a link to your website where customers can book an appointment with you online.

- Products: list the various products and offers that you would like to display to potential customers.

- Questions: be sure to list several frequently asked questions and be sure to answer the questions appropriately.

- Business Description: be sure to utilize all 750 characters that Google allows and describe all of your services and service areas in your business description.

- Business Category(ies): be sure to select a primary business category and select other categories that might apply. For example, Plumber should be your primary category and Water Restoration might be a second category if you provide those services.

- Services: list all of the services that you provide for each business category. Be sure to include all services that you perform no matter how big or small of a job.

- Service Territory: list the zip codes or sub-localities where you provide service to your customers. Be sure not to over exaggerate and only list the areas where you can actually serve customers.

- Photos: be sure to add a logo, cover photo, and images of job sites or your team. Continually adding photos to your GMB profile will help increase your rankings over time.

Completing all of these areas of your GMB profile shows Google that you care about your online presence and helps your map ranking.

Citations

A citation is a mention of your business on another website. When Google is able to identify your business (usually through an exact match of your NAP from your GMB listing), it adds credibility to your business. The reason these citations help to increase your ranking is that Google believes a fly0by-night plumbing business would not be able to get listed across the internet overnight. Only credible, established businesses have many citations. Using software like YEXT to help get your business listed on many websites is also a great idea.

As part of our service, we help our plumbers get properly listed in over 100 different online directories and websites to increase their business citations.

Posts & Social Media

Another important aspect of ranking higher on the Google map is to regularly post on your Google My Business (GMB)profile and on prominent social media platforms. Posting 3-4 times per week on your GMB profile shows Google that you are an active business. Google's algorithm will also read the text in your post captions and can help you rank for certain searches containing that text.

It is also very important to post regularly on prominent social media platforms such as Facebook, Instagram, and LinkedIn. Increasing your activity and online presence on these platforms has a direct impact on your Google rankings. As part of our service, we put out a daily custom social media post on all platforms including our clients' Google My Business profile.

Task Checklist

- Generate new Google reviews every day
- Respond to all Google reviews (good & bad)
- Fill out your GMB profile completely
- Add photos to your GMB regularly
- Get citations on other websites
- Post at least 3 times per week on your GMB
- Post at least 3 times per week on social media

Google Ads

Google Ads can be complex and expensive when not set up properly. On the other hand, Google Ads are an excellent way to scale your plumbing business when managed correctly. There are different types of Google Ads to run and each has its own individual benefits.
Do not get confused. We will show you how to run Google Ads profitably to grow your plumbing business into the future.

Google Local Services

Google Local Services ads, also known as Google Guaranteed ads appear at the very top of the page. These ads are not available in every market, but in the markets where they are, we suggest running them. These ads require that Google verify your business. You need to submit your insurance certificates, plumbing license, and each technician must pass a background check. The benefit of these ads is that you will only pay for actual customer phone calls. Google Local Services ads can be spotty, however. Google will rarely use your entire budget and the leads can be few and far between.

Google Local Services Set Up

To set up a Google Local Services campaign simply visit ads.google.com/localservices and start to complete your profile. After your account is approved, just let the ads run. Besides adjusting your budget, there is no manual maintenance required for this ad campaign.

You can log in to your ads account to view customer leads and even listen to the recorded phone calls. If you happen to get a spam call, you can even dispute the call with Google and they will reimburse you for the cost of that call.

Google Ads (PPC)

Formerly known as Google Ad Words, Google Pay-Per-Click ads are an excellent way to drive phone calls to your plumbing business. In a traditional PPC campaign, you can bid on certain keyword searches and have your ads appear in front of people with intent. I would need to write another book to explain all of the nuances of Google PPC ads, so I will be direct. Do not try and manage these ads by yourself. It is best to hire a Google Ads agency to manage your Google PPC ads or your could lose your shirt! We currently manage tens of thousands of dollars monthly in ads for plumbing clients and it is a full-time job!

Google Local Ads

A new ad campaign available on the Google Ads (PPC) platform is the local campaign. These campaigns pull information from your Google My Business listing and are great for local businesses looking to increase online traffic. These ads put more eyeballs on your Google My Business listing and can create significant call volume when set up properly. This is a campaign that you can manage on your own after you get some help setting it up correctly.

Your Website

Developing a quality website is critical to becoming a 21st-century plumber. Let's face it, everyone is online. They are shopping, banking, and searching for plumbers. When someone visits your website, its quality should represent the quality of your plumbing work. There are many do-it-yourself website platforms, like Wix and Square Space, where you can develop a quality website for a low cost. If you are interested in getting your website to rank higher on Google, then you may want to consider investing in a more robust website built on the WordPress platform. WordPress is more difficult to do yourself but has many more benefits when it comes to search engines. We have built many WordPress websites for our plumbing clients over the years, and there has been a noticeable increase in business for those clients with quality sites.

If you are interested in developing a new website for your plumbing business, I would encourage you to hire a web developer. If you have some creative skills, then you can easily build one on a platform such as Wix or Square Space.

Getting Your Website To Rank

Getting your website to rank higher on Google could be beneficial. If you have limited resources to invest in marketing, I personally would advise that you focus on your Google maps rankings and on running Google Ads. I believe those channels will produce more direct calls and revenue for your plumbing business.

Website ranking is always the last frontier I tackle with my clients. People searching for a plumber do not want to take the extra step of going to your website when they can read about you and call you from the map. If they have a larger project, such as a shower replacement, they might go to your website to check out your credibility. Otherwise, standard service calls are predominantly done from ads or the map.

With all that said, if you are still hard-headed about getting your site to rank higher. Here is how you do exactly just that...

Becoming the Authority

Google ranks websites that have authority on a certain subject. Remember, Google is trying to improve the searcher's experience, so displaying sites that give good information on the topic that is searched is their priority.

For your website to be the authority on a certain subject, such as tankless water heaters, the quality of your web content is very important. In addition, the structure and flow of that content are equally important.

On-Page SEO Structure

The first step to ranking your website higher on Google (Search Engine Optimization - SEO) is to structure the content on your site properly. After you have all of the information straight that you would like to include on your web pages. Your site should follow the structure outlined below:

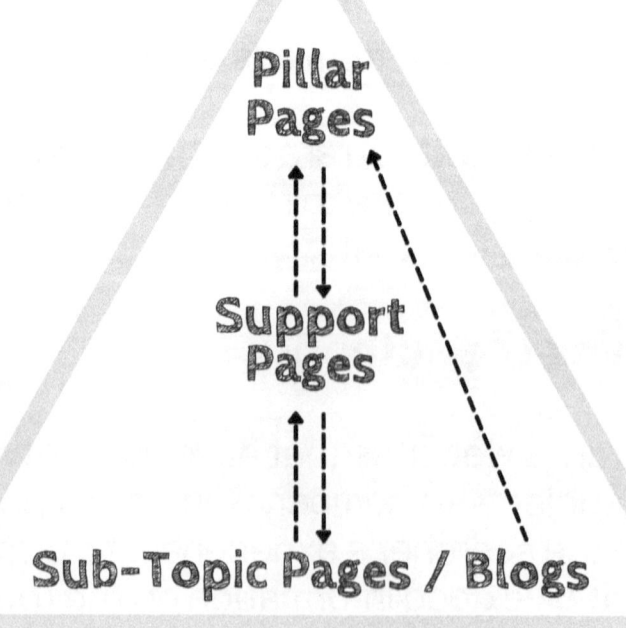

- Sub-topic pages and blogs should act as informative, easy-to-understand content for potential customers to read. If you are trying to rank your website for tankless water heaters, then a sub-topic page might offer the Top 10 Benefits of a Tankless Water Heater.

- Support pages help to support the content on the sub-topic pages and blogs by further explaining the missing details. For example, a support page for the sub-topic mentioned above might be how heating elements work within a tankless water heater to make it more energy-efficient. Sub-topic pages should offer internal links to support pages.

- Pillar pages should act as the ultimate authority on your topic and should include all of the nitty-gritty details. Support pages and pillar pages should interlink and sub-topic pages can also link directly to pillar pages.

If this confuses you, do not fret. Email me and I can walk you through it (sam@shwego.com).

Off-Page SEO

The next most important step to getting your website to rank higher is to get other credible websites to link to you as a reference.

When other credible websites link to your site as a reference, Google gives your site credit as more of an authority on that subject and you will rank higher. Here are some of the do's and don'ts when getting links from other sites:

Do

- get links from sites with similar content
- get links from sites with high authority
- get links from content that is relevant

Don't

- get links from non-credible sites
- link to yourself from sites you own
- get links from sites that are irrelevant

When other sites offer links to your content as a reference point, this is called a backlink. The more credible, relevant backlinks you can acquire to your site, the better.

There are many strategies to acquire backlinks, but the best way is manual outreach. Contacting sites that you want links from and offering content for them to link to is the slow and steady way to build credible links to your website so it will rank higher. Questions? sam@shwego.com.

Ongoing Marketing

Don't get stuck chasing success with new customers!

It is very important to master ongoing marketing, customer retention, and referrals to grow your plumbing business. Acquiring new customers is hard and expensive. Getting existing customers to come back when they have new problems or refer others to you should be a piece of cake! Many plumbers miss the ball here and are only focused on new customers. If you can master the follow-up and stay in front of your old customers, the chances of repeat business are much higher. The advantage of a repeat customer is that you do not have a customer acquisition cost. The largest plumbers in the country have been able to develop good reputations in their communities and have lots of repeat and referrals. Below, we will discuss how to develop this repeat business for yourself.

First and foremost, you must provide excellent plumbing service. Without that, you can kiss repeat business and referrals away. I will assume that you are an excellent plumber.

There are several methods you can implement to help with customer retention and referrals. I will outline each proven method that has worked for my plumbing clients below.

Email Marketing

- develop an email marketing list
- send a regular monthly email newsletter
- newsletter should include a coupon
- newsletter content should add value

Happy Calls

- follow up with customers via phone
- calls should be made 1-3 days after service
- check on their satisfaction and solve issues
- ask for online reviews and referrals

Property Managers

- prospect property managers in your areas
- these are great repeat customers
- develop a routine to contact these customers
- implement happy calls and emails to stay in front of your property managers

Some other methods we have implemented for our clients include text message marketing and evets/contests for existing customers.

Ongoing Campaign Set-Up

There are several out-of-the-x software programs you can use to set up an ongoing marketing campaign. We use a platform called Trumpia to perform email and text message marketing for our clients. We also use a program called MailChimp for ongoing monthly email newsletters.

These programs are inexpensive and relatively simple to set up an account. After you do so, you can create customer contact lists, campaigns, and rules to automate your ongoing marketing campaign.

If you want to keep it simple, just set up a monthly email newsletter to go out to all existing customers. Each time you get a new customer, be sure to add their email to your master contact list. This will allow you to stay in front of your customers so they might call you next time they have a plumbing issue.

If you need help setting up an email or text message marketing campaign, please contact me directly, and I would be more than happy to walk you through it (sam@shwego.com).

Technology

Becoming a 21st-century plumber means implementing the right technology and tools to automate the administrative part of your plumbing business.

There are several CRM programs available on the market that can help to streamline the mundane tasks of your plumber business, such as invoicing, project management, customer management, marketing, and more.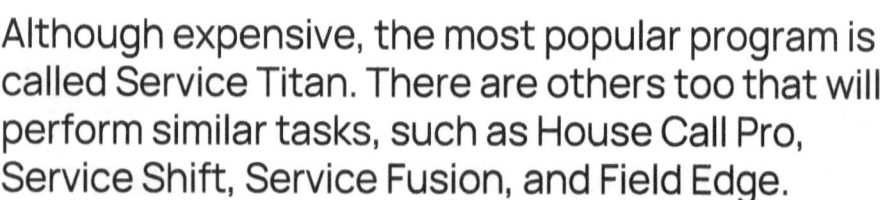

Although expensive, the most popular program is called Service Titan. There are others too that will perform similar tasks, such as House Call Pro, Service Shift, Service Fusion, and Field Edge.

All of these programs have a scheduling feature, allow you to invoice customers easily, collect payments, manage jobs, and integrate marketing campaigns.

Your CRM system should be the foundation that the rest of your plumbing business is built upon. If you need help choosing or implementing a CRM system, please contact me (sam@shwego.com).

Marketing Integration

Many of these plumbing software platforms come with great marketing integration tools. For example, in Service Titan, there is an automatic text messaging feature where you can text customers as soon as you close an invoice. This is a great way to provide customer service and even ask for referrals or reviews through the text.

Furthermore, some of these platforms allow you to assign jobs to various marketing channels. If you set them up properly, some can even do this automatically. This is a great tool for understanding your true return on investment when it comes to marketing dollars.

A Google Ads integration will allow you to see how much revenue was generated from your Google Ads budget, so you can truly manage your costs.

A good CRM will also allow you to run reports on customers so you can track the growth of your business, generate email marketing lists, and even wish customers happy birthday via email.

Most plumbers ignore the power of a good software program and are stuck in the past. Allowing your software to act as your marketing employee is a great advantage.

Next Steps...

If you have any questions about what you have read or are interested in getting help, I would encourage you to contact me. Thank you for reading this book, and good luck to you in the future! I wish you success!

Sam Puleo
Shwego Media
sam@shwego.com

~

THE END

www.ingramcontent.com/pod-product-compliance
Lightning Source LLC
Chambersburg PA
CBHW030542220526
45463CB00007B/2951